Contract Engineering in the Continental U.S.: Lessons Learned by a Technological Nomad

By Ken Rollins

Copyright 2012 Ken Rollins

Introduction

I have spent more than 20 years in the engineering field and nearly a decade of my career as a contract engineer.

During that time I have encountered a variety of corporate environments. I have spent time working for companies focused on doing all the stuff written on the hallway wall posters … innovation … integrity … teamwork. What a pleasure it is to walk into the office every day and have a spring in your step because you cannot wait to get to your desk and get started.

At other times, I have worked for drunks, the mentally disturbed and the grossly incompetent.

In this book, a "little black book" of sorts on contract engineering, I will try to relate what I have learned the Easy Way (from others and from books) and the Hard Way.

Copyright 2012 Ken Rollins

Definitions

Before we get started I need to define a few basic terms used herein.

I use the terms "contractor" and "temp" to refer to engineers working a professional, temporary, engineering-related job. The people working these kinds of jobs are not working directly for, or receiving benefits from, what is usually, a Fortune 500 "client" company.

Other terms you may hear while pursuing contracting work are "W-2 Contractor", "Corp-to-Corp Contractor", or "Consultant." The last 2 terms usually refer to someone who has created their own company (i.e., an LLC) for occasional or long-term use. More about that later.

A "W-2 contractor" and "contractor" are used synonymously in the succeeding parts of this book. A contractor will have a relationship with another intermediary party (referred to as a "job shop" or just "shop" typically) that provides the contractor with a paycheck drawn from funds provided by the client company. The shop will make payroll withdrawals for Social Security and Medicare as well as provide Unemployment Insurance and, many times, health and dental insurance plus a 401k plan of some kind.

A shop typically does not refer to itself as a "job shop." They may refer to themselves as technical consulting agencies or professional employment agencies. Whenever you talk to another contractor, they usually just use the term "shop."

Why Become a Contractor?

I cannot say this enough: contracting is not for the "faint of heart."

I have met former direct employees who became angry at their employer because of a lack of promotions and/or wage issues and decided to quit and become a contractor.

Because of outsourcing and other factors, the number of contracting opportunities has greatly diminished. The upheaval caused by frequently losing your job and changing job sites is also extremely hard on your family and on your own psyche.

Before you decide to become a contractor, it would be a good idea to get a firm grasp on what it is like to be a contractor by talking to one and to also check into how the current state of the U.S. economy is affecting contracting opportunities.

Generally, the U.S. economy is in flux to some extent during a Presidential election year. Wars and changes in corporate business models can also affect the contracting industry.

One of the other reasons some choose to become a contractor is out of necessity or desperation.

When you are in your 20's it takes about 2 to 3 months to find your first professional, direct job after graduating college. The length of your job searches changes as you grow older.

In your 30's, the time spent searching for a direct job falls more into the 3 to 6 months range and, after you are in your 40's, count on spending 6 to 9 months or more looking around for employment, depending on the demand for your particular skill set.

But if you are desperate and the bills are about to start piling up, contracting is a viable alternative to staying on the "direct employee" career path.

Copyright 2012 Ken Rollins

Depending on your job skills, a contracting job can be found within a month to 2 months. (More about "seasonal" influences later.)

It is your life, so it is your decision. I merely suggest you think about the consequences before you make your final decision.

Some scenarios where people relish and thrive at contracting are:

- You and your spouse have retirement income and you like to travel.
- You have a skill set that is in high demand (a good mix of hardware and software development is an example of this).
- You have a great social network locally and also possess out of state professional contacts.
- You have a large amount of savings (enough to live on for a year or more).
- You do not have children, a spouse, or a mortgage.
- You are a consummate Office Politician.

Qualifications

To be a contractor you need to have 3 or more years of engineering experience after having graduated with at least a 4 year degree.

I have met contractors who had nothing but a high school diploma and a lot of on-the-job training and experience. But they are the rare exception.

You should have a bachelor's degree in an engineering discipline (Electrical Engineering or Computer Science, for example) or one of the sciences, such as Physics or Chemistry.

I have met contractors who had degrees in Psychology and Geology. I have also met some who had master's degrees and even a few who had PhD's.

Local Versus Out of Town Jobs

People always prefer to sleep in their own beds, hug their spouses, and kiss their kids goodnight every night.

But every contractor I have ever met, no matter what their background, age, or intentions has done some out of town work. I have done a great deal of the latter so you will find several applicable sections in this book. Even if you are not planning to work out of town, it would be a good idea to read through those sections.

People who have great social networks can put off working out of town for quite a while, but eventually every local economy takes a hit, and so you hit the Road for an out of town "gig." It is not an insult to have to work out of town.

Working out of town has its own idiosyncrasies and I have included advice regarding some of those oddities with the intention of helping you deal with what is often a trying situation.

"Chaos Equals Cash"

It has been said this is a contractor's motto.

Think of your first day at work as parachuting into a battle zone.

Why so much chaos?

It's usually one of a handful of reasons:

- Poor people management.
- Poor project/schedule management.
- Undesirable job site location (such as New York in the winter or Yuma, Arizona in the summer).

Realize, you're a "hero" brought in to make up for someone else's screw ups. Also remember it is more than likely the job is just another temporary job that will not last the remainder of your career.

Never expect to work in the best of job climates. You will work under great supervision and in great circumstances sometimes, but in most situations the atmosphere is tense and the situation is not 100% "as advertised" during your job interview.

Take the cash, keep your resume updated, and move on when necessary.

Contract Lengths – Do Not Assume!

This is one of the biggest areas where things turn out to be something other than what you were told when you first accepted the job.

From the first paycheck, try to save at least 20% of each check. This is really tough to do whenever you are working out of town, maintaining 2 households, but I suggest this as a bare minimum amount. You just never know what is going to happen.

I also suggest you find a way to invest part of your money. I am not going to hand out investment advice, but find somewhere to put your money. As of 2012, there is at least one web-based investment service that will allow you to make very small investments – less than $10 per month – that I have used.

If your investment portfolio is large enough, you can sell off your investments between jobs for home emergencies or for "survival" money.

While you are working on a contract, some of your coworkers may say things like, "We're going to be here for years!" or "They can't possibly get rid of us. We know too much."

Do not believe it. This is why companies hire temp workers: so they can get rid of them and do not have to look bad to the direct employees, the direct employees' social network, or the rest of the community. Do not go out and lease a house or apartment based on the word of a job shop recruiter or coworker.

Try to get a copy of the project schedule after you start at your new job. Also, pay attention to the direct employees around you to get a better feel regarding how stable your employment situation is.

I would suggest avoiding the signing of any long-term rental agreement, whenever working out of town, until you have been on the job for 3 to 4 months. By then, you ought to know whom you can trust, be familiar with the schedule, know the "good" and "bad" areas of the city, and can make a more informed decision.

During the first 3 or 4 months, find an extended stay motel or, if you are so inclined, rent a room from someone.

Copyright 2012 Ken Rollins

Hourly Rates

First find out the local rates in your home town by asking an experienced contractor what the usual range is for your area. Do not be unprofessional and ask him or her how much they are being paid.

If you are going to be working out of town, look at the U.S. government per diem rates and do some math to figure out how much you should be getting if you plan to be working in another city or somewhere near a major city.

As of 2012, the expected rate for California is about 20% more an hour than the Dallas-Fort Worth rate. The Washington, D.C. area rate should be about 40% more an hour than the Dallas-Fort Worth area. Again, consult the U.S. Government per diem charts for current information.

Per Diem Split

Whenever you are working an out of town temp job, you are entitled to what is called a "per diem split."

This means that part of your paycheck goes untaxed. The daily amount is calculated using the U.S. Government per diem charts and the weekly amount is limited to a maximum of 50% of your base 40 hour paycheck.

To qualify for one, you must own or rent a permanent residence more than 50 miles from your job site.

You are allowed to collect this untaxed part of your paycheck for a maximum of one year.

If your shop is not going to give you a per diem split, you should expect a higher hourly rate that will equal the city's per diem rate.

If the shop is not going to give you a per diem split and they are not going to give you a higher hourly rate, I would suggest looking for another opportunity.

Job Shops – What are they?

I know of job shops that are run by 2 people out of a business park. Others are giant corporations finding temporary employees for even larger client corporations. Still others are someone's mom working out of her kitchen.

You just never know. The ones that offer per diem splits and benefits tend to be the larger, more established shops.

Currently, foreign recruiters are working for American job shops and also can have their own job shop in their own country, trying to find jobs for contractors in America.

Generally, the foreign recruiters are harder to understand, less professional by American standards, and very difficult to pin down on key issues such as per diem. You end up having to pin them down to a "yes" or "no" response using 2 or 3 emails and/or phone conversations.

A lot of shops will tell you that if your current assignment comes to an end to they will help you find your next job. Do not believe that.

Professionally speaking, you should let your current shop know if you are told your current assignment is coming to an end, but do not expect your shop to find you a new job ASAP.

Every day you are on your current assignment, the shop is collecting money. They want that regular paycheck and do not want to burn bridges with the client company at which you are working by helping you find a job elsewhere in an expedient manner.

In addition, if one's current shop does find your next assignment, they are most likely not going to find a job for you in a timely fashion so that you have continuous employment. Instead, whenever you have a job, save all the emails from recruiters that contact you while you are employed and contact them as soon as you know you are being let go from your current position.

Copyright 2012 Ken Rollins

Copyright 2012 Ken Rollins

Corp-to-Corp Contractors

I have no actual experience working as this type of contractor, but I will share what I know.

Basically, you are looking for a job and a shop says the client company will only take a corp-to-corp contractor.

The first rule, even if the rate is good (usually slightly more than you would expect as a W-2 contractor) and the job is offered to you, you only have to form your LLC if you take the job.

You can form an LLC via websites devoted to doing this sort of thing. You send for the forms, you fill them out and the forms are filed with your state of permanent residence. You pay for the forms and the state registration fees.

You will also pay a not-too-large monthly fee to your shop for liability insurance.

You will not receive a per diem split for this kind of job and when the job ends you will not be entitled to receive unemployment insurance payments because your corp-to-corp job ended.

In addition, since you are considered "self-employed," you will need to pay the IRS regular quarterly tax payments. Do not try to skip these payments as it is hard to find the money to re-pay "back" taxes.

Copyright 2012 Ken Rollins

Resume Websites to Use

Job search websites fall into two basic categories: free and pay-to-use.

I usually use at least 2 websites on which to place my resume when I first begin a job search. One charges an annual fee, but blasts my resume to over 150 job shops if I check the right box. If you have paid for such a service, the website should offer you a "history" of all mass emailing events and a phone number to call to make extra sure your resume was sent out as expected.

The other site I use is a free site.

For general job hunting, I used to use 2 free sites, one allegedly devoted to engineering jobs, but I noticed that one site was used frequently by potential employers and the allegedly more engineering-focused site was hardly used. I now use the less frequented site for only automated job searches.

If you are using a website and aren't getting any contacts from shops or headhunters, email or phone the people running the site. They can give you tips on how to use their site to draw employers to your resume.

While using one free site, I was told to check another box on a specific page and after doing so I started getting phone calls and emails based on the resume I had posted on their site!

Job Hunting Seasons

There are certain parts of the year that by the time they roll around you will prefer to have a nice long-term job.

The first season in which to avoid job loss is the fall. I have found myself in this situation more than once and I have a 50% success rate doing job hunting during this part of the year.

Even if a company's fiscal year is not based on the calendar year, they will still lay off employees during this time of the year and will only resume their hiring efforts after the end of the American holiday season.

The other time of the year during which it is hard to find a temp job is May and June. This is because all the new college graduates are job hunting and the summer interns are also being hired.

Purchase Orders and Onshoring Jobs

This is a job situation where a client company has told a shop to hire X amount of engineers with certain qualifications to do a series of engineering tasks.

In this situation, the client company has more of a relationship with the shop than it does with an individual contract employee.

It is a kind of "buy in bulk" approach to hiring contractors.

The strength of this situation is that the shop puts in place its own chain of command and management infrastructure, so you have more of a buffer between you and the client company.

The down side is that some companies may unexpectedly tell the job shop that the purchase order has to been scaled back by a certain dollar amount and so a group of people are let go without the client having a voice in the layoff process, resulting in high performing employees being let go just to make sure the nickels and dimes add up and a new purchase order can be signed.

Another scenario that arises out of this situation is one where the client company takes a long time renewing the purchase order (usually every quarter), resulting in you abruptly becoming unemployed for some indeterminate amount of time while the new purchase order is being approved.

Copyright 2012 Ken Rollins

"What was my start date again?"

I showed up early for a meeting at one client company and a manager and a first level supervisor were discussing the start date for a new contractor on their project.

The manager mentioned that they had originally told the contractor he could start on one date, but the manager thought he needed the contractor to start sooner. The manager was going to tell the shop the contractor needed to start sooner "or else," potentially damaging the contractor's relationship with his current employer because the new contractor would have to give less than 2 weeks notice at his old job to make it in time to his new job.

The odd thing was that I could swear they did the same thing to me.

Do not think that a manager at a major Fortune 500 company is above being petty and ruthless regarding your start date. You are just another temp to most of them.

Get it all in writing and keep the documentation (including emails) where you can find the agreed-to start date at a moment's notice.

Third Shift Jobs

I once took an engineering job that required long periods of third shift work. This sort of thing is really hard on your body, not to mention any family and friends you may have.

(FYI, it will help you sleep if you do not eat after midnight and you have a white noise generating device in your bedroom.)

I consulted my state's Unemployment Insurance department after the job ended to see if I could have turned the job down. I was told that if you believe accepting any job will negatively impact your health, you do not have to accept it and you do not have to mention "officially" that you turned down the job to Unemployment Insurance personnel.

I learned a lot of things during that job, but I would have preferred to have learned them during a more normal 1st/2nd shift engineering job!

Unscrupulous Job Shop Recruiters

I have had multiple run-ins and been warned about these kinds of recruiters. I am going to give examples and some basic guidelines to help deal with them.

I was offered a direct employee position at a company and had interviewed as well for a temp job at a different company via a phone interview. I was going to accept the direct job and decided to let the job shop recruiter know about my decision and gave him the name of my new employer, thinking I was being polite and keeping the job shop recruiter from wasting his time.

The temp job recruiter called into my new employer and talked them out of hiring me. Fortunately, I was able to find yet another temp job, but that experience taught me a big lesson.

Most recruiters do not want to burn any bridges with contractors and the recruiters behave ethically and celebrate the fact that you have found a job even if they did not place you at the new client company.

Even so, after you have found a new job, only tell recruiters where you have worked in the past and do not tell them where you are currently working if you have been at your current employer less than a month or have not started the new job yet. Just tell any job shop recruiters who contact you that you are no longer available.

Some recruiters have been known to tell prospective employers that a particular, newly hired contractor (let's call him Mr. Jones) was a poor performer when Mr. Jones worked for the recruiter's shop and that the recruiter has the resume of a better engineer (e.g., Mr. Smith) to present to the client company.

The whole story is a lie; Mr. Jones has never worked for the recruiter's shop. Mr. Jones is turned down by the client company and Mr. Smith gets his foot in the door. The recruiter collects a paycheck while Mr. Smith works at the client company.

Mr. Jones has to restart his job hunting activities.

Another type of recruiter I ran across allegedly scheduled a phone interview with a client company, but come the day of the interview, the client company's manager never called me.

To this day, I do not know what the recruiter was after. I just noted the recruiter's phone number in my cell phone so that I did not waste my time with his calls in the future.

I R & D Contracts

The acronym is pronounced, "eye rad" and stands for "Investigate, Research, and Develop."

This type of contract receives its funding via the back pocket of the client company or the United States Government instead of a specific program contract.

The common element between the 2 sources of funding is that the funding can dry up unexpectedly because of a lack of funds from the Government or a worried corporate CEO.

Usually when you are hired, a client company manager will tell you how long you can expect to be employed and you can usually count on them being correct. However, there are always exceptions and this specific kind of contract can shrink from to 2 years worth of work to only 4 months within a week after you arrive.

One contractor I heard about was told he had an I R & D job for 6 to 9 months – a typical contract length – but because he started at the end of the year, the Government told the client company that I R & D amounts would be much smaller the next calendar year. His 6 to 9 month job turned into a 6 week job.

If you are desperate, take the I R & D job.

But always ask if the contract is an I R & D contract during your interview with a prospective client company so that you do not get taken by surprise.

Work History Listed on Your Resume

If you have concerns about employers practicing age discrimination, you can limit your work experience to only the last 15 years.

As a full-time contractor, your resume is going to become quite lengthy. A little trimming will not hurt.

The important thing is to get an interview and then do your best to be hired.

Interview Types

Interviews fall into 2 basic categories.

The first is more "work history/experience" based.

For example, topics discussed might be how much experience you have with the C programming language and for what kinds of applications you have written software.

Another kind of interview is an "academic" style. This is like a verbal pop-quiz in college. It is favored by younger engineers and those that are inexperienced as hiring managers. More experienced managers are usually familiar with ways of determining if you are lying (which I will cover in a later section) and do not feel the need to give you a "vocabulary" test.

Younger engineers, lacking the familial and financial responsibilities of older more experienced engineers, have more time to keep up with the latest technological issues and therefore do not possess the people skills or the experience to tell when they are interviewing a talented, dedicated, experienced engineer. So they fall back to what they know: asking questions like their professors in college.

Academic interviews can also include "tests." I have encountered 2 of these types of interviews in my 20+ years of engineering and I failed both of them. I usually just look up any "academic" type issues on the web once I start a job and do not worry about memorizing factoids for an interview situation.

I encountered a contractor once, the suit-and-tie type, who undoubtedly excelled at academic style interviews. However, he did not have a "head" for engineering, so he usually only lasted 2 months on a job.

Finally, whether you have a background in hardware or software, if you have an interview related to software development on any level, you may be questioned about processors on which your code executed and even how many lines of code you have written in a particular programming language.

Supervisory Jobs

One of the most important things to remember with this kind of job: do not expect to be friends with those whom you supervise.

The reason for this is that it makes you more susceptible to issues like sexual harassment claims or employees lying about their work experience because you have your guard down.

In these 2 scenarios, the other person's agenda is to get rid of you. You letting your guard down (as one does with friends) enables them to carry off their schemes.

To further prevent any negative experiences as a supervisor, if possible, do not sit in the same cubicle with anyone you supervise. People need a chance to "vent" about their boss to let off steam.

This last tactic will also prevent anyone gaining information that they could use to stab you in the back.

To this end, avoid web surfing while your staff are in the office.

Also, know that any employee who has lied their way into an engineering environment will eventually be found out and just about the time they are, they will be especially vicious.

Do your best to not give an unscrupulous employee any "intel" they can use to hurt you before they get shown the "exit."

Copyright 2012 Ken Rollins

Ethics Training and a Lack Thereof

A lack of ethics training for contractors leads to an abundance of unethical contractors in the workforce.

These days some kind of background check is common for even temporary workers like contract engineers. If you find yourself working for a company that does not provide ethics training for contractors, fall back to the highest ethics standards you have been taught elsewhere and do not be afraid to be assertive about asking questions regarding intellectual property standards and security issues.

You do not want to accidentally do something that will affect your ability to find employment in the future.

In such cases where ethics training is lacking, expect to encounter contractors who have an "anything goes" attitude. Collect your paycheck and concentrate on producing the highest quality work you can and do not lower your standards.

Time Charging

This relates back to ethics training.

As a contractor you are to be paid for every hour you work. You do not give a company free time while present in the office. Ever.

Doing research on your own time, after hours and away from the office is acceptable.

Having said that, I know of at least one contractor who would give client companies a free half hour here or there to make up for any time he spent chatting with coworkers. He felt it helped his professional reputation and gave his conscience peace of mind.

Be aware that not charging for your time is illegal.

If a company is paying you so well that you do not need to be paid for your overtime, it gives that company a competitive edge.

In addition, mischarging your time causes false schedule metrics to be generated for current and future contracts.

Finally, when one contractor charges for all the hours he or she has worked and a nearby contractor does not, guess which one (mistakenly) looks like the better engineer? This is not a way to create positive, long-term relationships with your peers.

Copyright 2012 Ken Rollins

Overtime

This topic usually comes up during an interview, but sometimes one or both parties forget to bring it up.

Most situations fall into one of 2 categories:

1. You are allowed to work/charge for overtime.
2. You are not allowed to work any overtime.

I have been in situations where working overtime up front was not only allowed, but encouraged.

I have also been in situations where there was an unspoken limit on how much overtime I could work and ended up being chewed out by a manager because no one had told me about the hourly limit.

The best way to cover yourself is to verify you can work overtime once you show up at a work site and ask about any time charging limits.

Most companies pay only "time" for overtime, but I have also been in situations where I was paid time and a half. This information is usually included with your employment paperwork you received from your job shop; therefore, it is a good idea to always keep the paperwork associated with your current job assignment close at hand.

Copyright 2012 Ken Rollins

The Honeymoon Period

This period of time consists of the first month to a month and a half you are on a job.

During this time, management is deciding if you are an "okay" fit for your job, a spectacular fit, or someone they would like to let go.

Do not worry about that last part. People who are let go in the first month or so are usually people who have lied on their resume, or have drug and alcohol issues.

When given your first task at your new job, swing for the fences! Do the task as well and as quickly as you can. This will establish trust between you and your employer and you will be able to settle into your new job faster because your mind is at ease because of the trust that has been established.

The 6 Month Trust Rule

This is a big life/professional lesson to learn.

As elsewhere in life, with contracting you are constantly going to be running into strangers. Within the first 6 months you work with a peer or manager, you figure out if they are a saint, a habitual liar, a drunk, a cheat, or some other kind of "disturbed" person.

Be aware that the best con artists working in an engineering environment can fool people up to 6 months and as you reach the 6 month point they can become very vicious and lethal to your professional reputation with the client company.

In summary, watch your back for the first 6 months and limit your socializing to only group, company-initiated activities if possible.

Copyright 2012 Ken Rollins

Evil Contractors

I have touched on this in other sections, but I thought I would lay it out plainly here.

You will meet a lot of fellow contractors who are just trying to make the best of the chaos wherein contractors work.

You will also meet:

- Liars.
- Thieves (stealing personal property).
- Timesheet frauds (stealing from the client company).
- Unfaithful spouses/significant others.
- Blackmailers.
- Racists.
- Male chauvinists.
- Web usage addicts.
- Emotionally and mentally disturbed individuals.

So if you did not know about or expect this before, you will now.

Prejudice – How It Affects You

Before the year 2000 and the advent of outsourcing, there were plenty of contract engineers who would tell direct employees they were stupid for working as directs.

These contractors would also frequently brag about how much money they made.

(I try to dress "down" when I am working in the office to prevent any jealousy. And I tell people that I make good money when I am employed, but I make zero dollars whenever I am unemployed. This usually alleviates any issues people might have with me.)

Because of these earlier contractors, there are some managers (even the one that hired you!) and other direct employees who despise contractors.

These direct employees see contractors as overpaid, disposable employees and cannot wait to see you leave the building with your belongings in a box.

Other client companies will say over and over again that their company treats contractors just like regular, direct employees.

No matter what level of friendliness you encounter, you will eventually encounter the "You're just a contractor" attitude at any client company.

Keep that in mind and you will not be disappointed.

Copyright 2012 Ken Rollins

Office Politics 101

If you are a consummate office politician, you are probably not going to have any "political" issues as a contractor.

For the rest of the engineers reading this, I have the following tips.

My 2 political behavior modes are:

- Be Friendly
- Be Quiet

You really cannot go wrong choosing one of these in any given situation. Even so, you wonder what people think of you sometimes.

When discussing something allegedly factual, if a person's voice pitch goes up, they are probably lying and/or hiding something.

In addition, if someone looks to their right hand side while speaking to you, they are most likely telling you the truth or recalling an actual event in their past.

If they look to the left, they are either lying or visualizing a scenario related to the conversation you are having.

If they stare hard into your eyes, they are likely trying to convince you they are telling you the truth because you do not seem to trust them and, in this case, the other person may be lying. The other person's piercing stare may also indicate they just want very much for you to believe them.

In addition, sometimes managers or coworkers that are familiar with body language basics will accuse you of something just to observe your reaction to see if you are lying. For example, they might say, "Rollins, I want the truth this time. Not your usual lies."

If you want to learn more about "reading" people, there are plenty of books devoted to the subject. Look for books about "reading people", "body language" or NLP (Neuro-linguistic Programming) for example.

Copyright 2012 Ken Rollins

Giving Notice

When giving your notice, always give at least 2 weeks notice. As an older engineer told me in my early 20's, "You should always try to avoid burning bridges."

So why should you be giving notice?

A couple of good reasons are:

1. You looked at an official copy of the project schedule and your current project is ending soon and you have not heard of any future projects that will be staffing soon. Therefore, you do not expect to be moved to a new project when your current program ends.
2. A client company manager gives you and other contractors a heads up that the client company will be letting you go soon.

With regard to the last scenario, the manager will give you a time frame in good faith. However, that can change. It is a very "fluid" situation.

The fluidity can be caused by funding issues arising from the client company's customer, or because some project manager wants to prove he is a "tough guy" and not afraid to drop the axe on a bunch of employees that are "just contractors."

The 3 Month Adjustment Rule

This rule applies to any job, but especially to out of town jobs.

To acclimate to a new city and a new job takes about 3 months or so.

If living out of town, it is at about this time that you will start referring to your new motel or apartment as "home," though technically it is not.

If you continue contracting for more than 5 years, you will become very aware of this pattern. Do not share this Rule with your boss; they expect you to be settled in and fully productive within a month or so and they may never have worked out of town or changed jobs as much as you.

A Nightmare Scenario

I have heard of only one instance where a contractor signed their new hire paperwork, did their drug test and set off to an out of town job only to have the client company cancel their contract before the contractor ever set foot in the door.

Bearing this in mind, if you are receiving Unemployment Insurance payments, I suggest not canceling them until you receive your first paycheck from your new job. Just to be safe.

TIPS FOR WORKING AWAY FROM HOME

Copyright 2012 Ken Rollins

Home Care

I use family and friends to help take care of my home when I am working out of town.

"Home checker" services are available via the web and as a homeowner I have had a lot of lawn care companies leave their cards on the door.

In addition, some of these home care services offer to occasionally start any vehicles you have stored on your property.

If you do not have access to home care companies, you will have to either schedule more frequent trips home (spending more money than one would prefer) or offer to pay your friends and neighbors to look out for your home as I do (spending a much smaller amount of money that you would have on flights home).

Mail

    The rule of thumb here is to forward your mail via the internet and cancel your mail being forwarded in person. Otherwise, things begin to go in circles.

    With regard to deciding to have your mail forwarded to a "Mail Box" store's mail box versus a regular U.S. Post Office box, go with the USPS. It is more difficult to cancel your mail being forwarded to a "Mail Box" store than to a U.S. Post Office.

Transporting Your Belongings

The methods in this category I have seen used most often are:

1. Mail your belongings to your new work site and/or back to your home when your assignment is over.
2. Rent a trailer to convey your belongings to and from your work site.
3. Use your normal vehicle to move your belongings around the country.

Each night of travel take the most expensive and private belongings into your hotel. As long as you stay in a hotel priced in the area's median price range near a major highway, your remaining belongings should be safe in your car or trailer overnight. I myself have never had any issues with theft or vehicle damage.

Also, if you do not have some sort of navigational system, buy one. Some states are just not well organized about traffic control in and around detours and a good GPS/navigation system can keep your blood pressure low in these situations.

Loneliness

I know … it hurts being away from your loved ones and your home.

To cope with this, bring reminders from home with you on your "away" jobs.

Anything will do. For example:

- A gift from your spouse or kids.
- A photo of family and/or friends
- A clock set to your "normal" time zone.

With regard to photographs, you can email these to yourself or carry a thumb drive and print them out at a local store in your new city. Make as many copies as you need to make you smile when you look around your hotel room or apartment. Also, do not forget to print some for work as well.

Family

My kids are old enough to know that I only work out of town so I can get a paycheck.

But at first, it was tough on all of us. Lots of tears were shed.

If possible, during your 3rd or 4th week at your new work site, schedule a trip back home.

I also make a habit of not only calling my kids regularly, but emailing them frequently and occasionally buying them spontaneous gifts online.

It takes effort to make any relationship work; even one with your kids.

I have had to cancel trips home, so buying "trip insurance" is something you may want to look into just in case.

Reasons for my canceling my trips back home were:

- Illness.
- Being informed my job would be ending in a month or less.

My trips home normally occur at 2 or 3 month intervals. Families with more than one income may want to schedule trips more frequently due to a larger family budget.

I have heard of semi-retired contractors that bought an RV, a small vehicle to tow behind the RV, and that lived out of the RV. Each day, the contractor would be dropped off while their significant other took the day off and had use of their one and only available vehicle.

In addition, I have heard of families where both parents were contract engineers and either mom or dad would take time off between assignments while their spouse worked so that one parent was at home all the time and the other parent was gainfully employed.

Copyright 2012 Ken Rollins

Population Factors

If you have not lived in cities of varying sizes, I offer you this advice.

Cities whose metropolitan area populations exceed one million are usually more singles-oriented, faster paced, and incidents of rudeness are frequent.

Cities whose population is under one million tend to be more family oriented, friendlier, and slower paced and you may find your fellow employees more prone to suggesting you become a direct employee at the client company.

In this last situation, always take a "direct" job offer or suggestion to become a direct as a compliment and adopt an attitude of gratitude whether you pursue the opportunity or not.

Health Issues

If you have any food allergies and are used to eating at a limited number of "safe" restaurants in your home city, you will have to be more careful during your travels. Otherwise, your body will remind you that you have forgotten to be more cautious.

If you have any kind of sinus allergies, it would be good to check what the normal air quality issues are where you are planning to work. Some precautions to take include air purifiers, antihistamines, nose strips and, if bad enough, wearing a painter's mask while sleeping.

One area I worked in had a particularly toxic plant that bloomed for 2 or more months each year. This particular plant's pollen made about 95% of the employees ill, including myself.

Whether working out of town or not, try to keep your health issues to yourself if possible. Health issues can be perceived as threats to individual productivity and to the highly expensive project on which you have been hired to work.

Finding a Doctor

If you regularly take prescription medications, it is always good to leave your home town with a few refills "on file" at a national pharmacy chain if possible.

If you have left town with less than a month's worth of any medication you take regularly, an Urgent Care can provide 1 or 2 months of refills. Make it a priority to find a family practice doctor that can take care of your healthcare needs as soon as possible though.

Be aware that some doctors do not take new patients who do not have health insurance or any new patients at all, so do not put off finding a doctor in your new city until you are ill.

Copyright 2012 Ken Rollins

Meet the Neighbors – OR NOT

Traveling and working out of town, you run into some "interesting" people.

Some I have encountered include shouters, drunks, home invaders, prostitutes, drug users, and the physically abusive.

To help with criminal matters, make sure you call 911 from your cell phone. Most hotels and extended stay motels are alerted if you call the police from your room. You do not want the "bad guys" knowing you were the one who "ratted" on them because they see hotel personnel suddenly flock to your room before the police arrive.

To help with noise issues, consider buying a white noise generator like an air purifier and quality, but disposable, ear plugs. If you use ear plugs, be careful to read all the instructions and warnings on the box to prevent permanent damage to your hearing.

Credit Cards – Re-Issuance Issues

Make sure the institution that provides your credit or debit card has your correct mailing address in your new city on file. If one of your cards expires, the new one will be returned to the bank or credit union that mailed it if it is sent to your permanent residence while you are having your U.S. mail forwarded.

Also you will more than likely become much more acquainted with your preferred financial institution's 800 numbers, email addresses, web site, and even their personnel while you work out of town. It is just a fact of life.

Vehicle Registration and Drivers Licenses

Most cities and states expect newcomers to re-register their vehicle and obtain a valid state drivers license if you are planning to live there more than 30 days.

Having said that, I have found that enforcement of these laws varies from situation to situation.

Network with other contractors in your work area, or just listen in on such conversations to see what the local customs are where you work.

Some places I have worked, the cops regularly patrol parking lots, noting license tags and re-checking them periodically and issuing citations if one does not have a valid in-state tag.

Other client companies hire so many temporary workers that (probably) for the benefit of the local economy, the cops do not care as long as you are a temporary employee. The police do not want to discourage strangers coming to town to work and spending their money even if the newcomers are there temporarily.

The Cost of Settling In/Leaving Things Behind

Some of the things I buy each time I move to a new city because of work are:

- A printer/copier.
- A space heater.
- An air purifier.
- Extra toilet paper.
- Paper towels.
- Facial tissue.
- Plastic utensils.
- Plastic cups.
- A TV tray.
- A laptop cooler for my laptop.
- Speakers for my laptop.
- A step stool.
- Hand soap.

Things I usually leave behind when I leave town (the people who work at your apartment or extended stay motel usually cannot wait to get their hands on these, by the way):

- The printer/copier.
- The space heater.
- The air purifier.
- Toilet paper.
- Facial tissue.
- Paper towels.
- The TV tray.
- The laptop cooler.
- The laptop speakers.
- The step stool.
- Hand soap.

Do you see the pattern? If you try moving this stuff over and over again, it just gets tiresome. But if you do not believe me, you can try it yourself.

FYI, as a rule, I put the space heater away in a cupboard when I am not using it. Motel and apartment personnel may not approve of you possessing such items. I always err on the side of caution.

Working Outside the U.S.

I hope I never have to leave my native country, but if there is one thing I have learned doing contracting, it is never say "never."

What I have heard is that you are usually assigned a special compound or living quarters in most Middle Eastern countries. If you do not obey their local customs, you will be deported ASAP. If you have a significant other traveling with you, they will be deported as well.

AFTERWORD

I am Ken Rollins and I approve this material.

Seriously, I hope what I have written has helped you. As an engineer, I try to always put forth an effort to better the Human Condition, so I hope this book has improved yours.

See you on the Road!

K.R.

www.ingramcontent.com/pod-product-compliance
Lightning Source LLC
Chambersburg PA
CBHW021040180526
45163CB00005B/2218